ELECTRICAL DESIGNING AND DRAFTING

JAPHET K. LAGAT

DEDICATION

To my Parents,

For always loving and supporting me.

To my brothers and sisters.

What…a family

CONTENTS

ACKNOWLEDGMENTS

Writing a book is harder than I thought and more rewarding than I could have ever imagined. None of this would have been possible without support of my parents, *Mr. & Mrs. Philip K. Kimeto*. They stood by me during every struggle and all my successes. That is true parenthood.

They taught me discipline, tough love, manners, respect, and so much more that has helped me succeed in life. I truly have no idea where I'd be if they hadn't given me a roof over my head or became the father/mother whom I desperately needed at that age.

To my Siblings. To Sis Jane: for always being the person I could turn to during those dark and desperate years. She sustained me in ways that I never knew that I needed. To my little brothers, Elisha, Emanuel and sisters, Brigid and Mary: thank you for letting me know that you had nothing but great memories of me. So thankful to have you in my life. Finally, to all those who have been a part of my getting there: Benson, Jackson, Jasper and Paul.

BASICS OF ELECTRICAL ENGINEERING

Father of Electricity

- "William Gilbert" and "Michael Faraday" is known as the father of electricity.

Electron Theory

- Electron is the elementary particle carrying a unit charge of negative electricity.

- This electron theory was discovered in the early 1900's by "Lorentz" and "Drude".

- It states that when potential difference is applied to a conductor the electron present in conductor will possess some kinetic energy and this Kinetic energy will start moving in a definite direction with a speed of light creating electrical pressure.

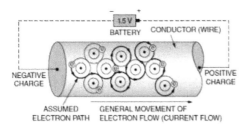

Voltage

- Voltage is also called Electromotive Force (EMF), is a quantitative expression of the potential difference or power distribution in charge between two points in an electrical field.
- The greater the voltage, the greater the flow of electrical current through a conducting or semi-conducting medium for a given resistance to the flow.
  ### Units: Volts (V)
- Measured by voltmeter, Multimeter, Potential Transformer (or) Voltage Transformer.

Current

- An electric current is a flow of electric charge. In electric circuits this charge is often carried by moving electrons in a wire. In simple words we can say the rate of flow of electrons.
  ### Units: Amperes (A)
- Measured by Ammeter, Clamp meter, Multimeter, Current Transformer.

Conductor

- It is a material that allows the free flow of electrical current in one or more direction.

Example: Copper, Aluminium, Silver, GI, Gold, Platinum etc. (All Metals).

Note: Platinum is best among all the metals.

Insulators

- Insulator does not allow the free flow of electrons. It is opposite of the conductor.
  Example: Paper, Rubber, Plastic, Teflon, PVC, Glass, Wood etc.

Semi-Conductor

- It is a combination of insulator and conductor.
- A semi-conductor is a substance usually a solid chemical element that can conduct electricity under some conditions but not others. Making it a good medium for the control of electrical current.
- In short whenever current is giving, its acts as a conductor. When current is not giving, its acts as an insulator.
  Example: SCR, Thyristor, Transistor etc. (All Power Switches).

Types of Electricity

(i) **Static Electricity:** Static Electricity is when electrical charges developed on the surface of the materials. It is usually caused by rubbing materials together because of friction.

(ii) **Dynamic Electricity:** Dynamic Electricity is the flow of electric charges through a conductor. Current is rate of flow of electron. It is produced by moving electrons.

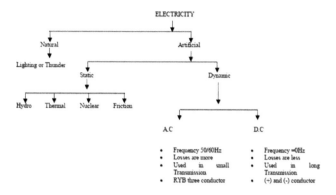

Power System

- Generating + Transmission + Distribution + Utilization

    (i) Generation
        Generated voltages are 11kV to 400kV (or) 440kV

    (ii) Transmission
        It is done by means of HVAC and HVDC lines

(iii)   Distribution
It is done by means of sub-stations
Total Expenditure of power system in percentage
Generation 40%, Transmission 20%, Utilization and Distribution 40%

Flow Chart

11kV – Generation
↓
440kV - Step Up
↓
220kV - Step Down
↓
132kV - Step Down
↓
66kV - Step Down
↓
33kV - Step Down
↓
22kV - Step Down
↓
11kV – Distribution
↓

1φ   3φ

Power Flow Analysis

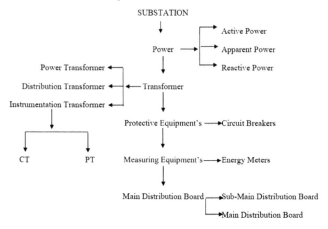

Types of Electrical Loads

(i)   **Resistive Loads:** It is a physical quantity which opposes the free flow of electron. Resistive loads are typically used to convert current into forms of energy such as heat. V and I are in phase for a purely resistive load, so power factor is Unity. No device is purely resistive in nature.
Units: Ohms (Ω).

Phase Diagram of Resistive Loads

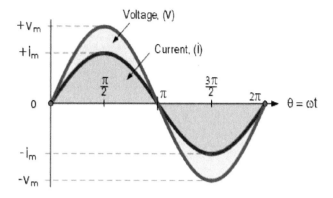

Phase angle between voltage and current is unity.

$$P.F = \cos \phi$$

$$Phase\ angle = \phi = 0^o$$

$$\cos \phi = \cos 0^o = 1$$

For purely resistive loads, $\cos \phi = 1$

(ii) **Inductive Load:** It opposes sudden change in current and stores current in the form of magnetic field. Electromagnetic fields are the key to inductive loads.

With an inductive load the current wave form is lagging the voltage wave form.

Inductive load is a load which pulls large amount of current or an inrush current.

Units: Henry

Phase Diagram of Inductive Load

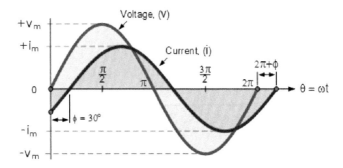

$P.F = \cos \phi$

$Phase\ angle = \phi = 90^o$

$\cos \phi = \cos 90^o = 0$

For purely Inductive loads, $\cos \phi = 0$

NB: Lagging angle (ɸ), varying with the device used.

Example of Inductive loads: Fans, pumps, Motors, Welding Machines etc.

    (iii)   **Capacitive Loads:** It opposes sudden change in voltage and stores energy in the form of current. The capacitor load has a current wave form which is leading the voltage wave form, therefore the voltage peak and current peak are not in phase. To enhance the power factor, we use capacitor banks.

Phase Diagram of capacitive Load

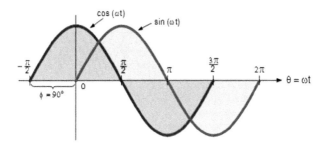

$P.F = \cos\phi$

$Phase\ angle = \phi = -90^o$

$\cos\phi = \cos -90^o = 0$

Example of Capacitive Loads: Capacitor banks, Synchronous condenser etc.

Different Voltage Level Present Across the Globe.

Low Voltages	220V, 230V, 415V, 440V
Normal Voltages	3.0kV, 6.6kV, 13.2Kv
High Voltages	18.2kV, 22.0kV, 33.0kV
Extra High Voltages	66kV, 132kV
Modern Extra High Voltages	220kV, 400kV, 440kV
Ultra-Extra High Voltages	765kV

Note:

- In 90% projects, 11kV is generally used as point of supply (POS).

BASIC FORMULAE

There are basically three types of Electrical Power

1. Active Power / True Power / Real Power

Definition: The amount of power (useful) we are consuming

Units: Watts / kilo Watts / Mega Watts

$P = \sqrt{3} x V x I x \cos \phi \dots \dots \dots \dots \dots (3\phi, 415V)$

$P = V x I x \cos \phi \dots \dots \dots \dots \dots \dots (1\phi, 230V)$

$\cos \phi = 0.8$

Rule: Up to 5kW, we consider the power as 1ϕ, 230V

Above 5kW, we consider the power as 3ϕ, 415V

Up to 6HP, we consider the power as 1ϕ, 230V.

1HP = 746W

2. Apparent Power

Definition: The equipment's which are giving (feeding) as the supply i.e., Transformer, Generator, UPS, Inverter, Alternator etc.

Units: Volt Ampere / kilo Volt Ampere / Mega Volt Ampere

$P = \sqrt{3} x V x I \dots \dots \dots \dots \dots \dots \dots \dots (3\phi, 415V)$

$P = V x I \dots \dots \dots \dots \dots \dots \dots \dots \dots (1\phi, 230V)$

11

Rule: Up to 25kVA, we consider the power as (1ф, 230V)

Above 25kVA, we consider the power as (3ф, 415V)

3. Reactive Power

Definition: The equipment's which are used to maintain the quality of power, i.e., capacitor.

Units: Volt Ampere Reactance / kilo Volt Ampere Reactance

$$P = \sqrt{3} x V x I \sin \phi \dots \dots \dots \dots \dots \dots \dots (3\phi, 415V)$$
$$\sin \phi = 0.6$$

i.e.

$$\sin \phi = \sqrt{1 - \cos^2 \phi}$$

$$= \sqrt{1 - (0.8)^2}$$

$$= \sqrt{1 - 0.64}$$

$$= \sqrt{0.36}$$

$$= 0.6$$

Rule: kVAR is used only for 3ф, so formula remains same for all.

NB: By looking at the units in the devices, we can differentiate either active / Apparent / Reactive.

Examples

Solve the following;

   (i) Find out the current (I) for all the given loads?

400W, 2000kW, 65kVAR, 125 kVA, 1300W, 40HP, 14kW, 5kVAR, 30kVA, 375kVA, 5000W, 15HP, 50HP.

Solution;

   1. 400W, It's a (1ф, 230V)

We know that; $P = VxI \cos ф$

   P = 400W, V = 230V, $\cos ф = 0.8$

   ∴ *I = ?*

$$I = \frac{P}{V \cos ф} = \frac{400}{230x0.8} = 2.17A$$

   2. 2000kW, It's a (3ф, 415V)

We know that; $P = \sqrt{3}xVxI \cos ф$

   P = 2000kW, V = 415V, $\cos ф = 0.8$

   ∴ *I = ?*

$$I = \frac{P}{\sqrt{3}xVx \cos ф} = \frac{2000 \ x \ 1000}{\sqrt{3}x415x0.8} =$$

$3478A$

   3. 65kVAR, It's a (3ф, 415V)

We know that; $P = \sqrt{3}xVxI \sin ф$

   P = 65kVAR, V = 415V, $\sin ф = 0.6$

   ∴ *I = ?*

$$I = \frac{P}{\sqrt{3}xVx \sin ф} = \frac{65 \ x \ 1000}{\sqrt{3}x415x0.6} =$$

$150.7A$

4. *125kVA, It's a (3φ, 415V)*

We know that; $P = \sqrt{3} x V x I$

$P = 125kVA, V = 415V$

∴ $I = ?$

$$I = \frac{P}{\sqrt{3}xV} = \frac{125 \; x \; 1000}{\sqrt{3}x415} = 173.9A$$

5. *1300W, It's a (1φ, 230V)*

We know that; $P = V x I x \cos φ$

$P = 1300W, V = 230V, \cos φ = 0.8$

∴ $I = ?$

$$I = \frac{P}{V \cos φ} = \frac{1300}{230x0.8} = 7.06A$$

6. *40HP, It's a (3φ, 415V)*

1HP = 746W

∴ *40HP =40 x 746W*

=29,840W or 29.84 kW

We know that; $P = \sqrt{3} x V x I \cos φ$

$P = 29.84kW, V = 415V, \cos φ = 0.8$

∴ $I = ?$

$$I = \frac{P}{\sqrt{3} \; xVx \cos φ} = \frac{29.84 \; x \; 1000}{\sqrt{3}x415x0.8} =$$

51.89A

7. *14kW, It's a (3φ, 415V)*

We know that; $P = \sqrt{3} x V x I \cos φ$

$P = 14kW, V = 415V, \cos φ = 0.8$

∴ $I = ?$

$$I = \frac{P}{\sqrt{3} \; x V x \cos \phi} \qquad = \frac{14 \; x \; 1000}{\sqrt{3} x 415 x 0.8} =$$

24.34A

    8.   *5kVAR, It's a (3ϕ, 415V)*

We know that; $P = \sqrt{3} x V x I \sin \phi$

        P = 5kVAR, V = 415V, $\sin \phi = 0.6$

        ∴ $I = ?$

$$I = \frac{P}{\sqrt{3} x V x \sin \phi} \qquad = \frac{5 \; x \; 1000}{\sqrt{3} x 415 x 0.6} =$$

11.59A

    9.   *30kVA, It's a (3ϕ, 415V)*

We know that; $P = \sqrt{3} x V x I$

        P = 30kVA, V = 415V

        ∴ $I = ?$

$$I = \frac{P}{\sqrt{3} x V} = \frac{30 \; x \; 1000}{\sqrt{3} x 415} = 41.73A$$

    10. *375kVA, It's a (3ϕ, 415V)*

We know that; $P = \sqrt{3} x V x I$

        P = 375kVA, V = 415V

        ∴ $I = ?$

$$I = \frac{P}{\sqrt{3} x V} = \frac{375 \; x \; 1000}{\sqrt{3} x 415} = 521.7A$$

    11. *5000W, It's a (1ϕ, 230V)*

We know that; $P = V x I \cos \phi$

        P = 5000W, V = 230V, $\cos \phi = 0.8$

        ∴ $I = ?$

$$I = \frac{P}{V \cos \phi} = \frac{5000}{230 x 0.8} = 27.17A$$

12. *15HP, It's a (3φ, 415V)*

1HP = 746W

∴ *15HP =15 x 746W*

=11,190W or 11.19kW

We know that; $P = \sqrt{3}xVxI \cos \phi$

$P = 11.19kW$, $V = 415V$, $\cos \phi = 0.8$

∴ $I = ?$

$$I = \frac{P}{\sqrt{3}\,xVx \cos \phi} \qquad = \frac{11.19\ x\ 1000}{\sqrt{3}x415x0.8} =$$

19.45A

13. *50HP, It's a (3φ, 415V)*

1HP = 746W

∴ *50HP =50 x 746W*

=37,300W or 37.3kW

We know that; $P = \sqrt{3}xVxI \cos \phi$

$P = 37.3kW$, $V = 415V$, $\cos \phi = 0.8$

∴ $I = ?$

$$I = \frac{P}{\sqrt{3}\,xVx \cos \phi} \qquad = \frac{37.3\ x\ 1000}{\sqrt{3}x415x0.8} =$$

64.86A

Assignment;

Find out the current (I) for the given loads;

550W, 65kVAR, 18kVA, 70HP, 10kW, 2000kW, 75kVAR, 128kVA, 75kVA, 3000W, 6HP, 100HP.

      (ii)   Find out the power (P) in terms of kW and kVA?

1000A, 125A, 15A.

Rules:

Active Power

To calculate in terms of W or kW, we consider the factor of 5kW, to find the currents, for both 1φ and 3φ
i.e.
Up to 27A………... (1φ, 230V)
Above 27A………...(3φ, 415V)

Apparent Power

To calculate in terms of VA or kVA, we consider the factor of 25kVA, to find the currents, for both 1φ and 3φ

{25kVAcosφ / 27kVAcosφ}, φ=0.8
i.e.

Up to 18A………...(1ф, 230V)
Above 18A………..(3ф, 415V)

Solutions:

1. *1000A*

kW =? *And* kVA =?

a) *For Active power in W or kW*

We know that; $P = \sqrt{3} x V x I \cos ф$
1000A >27A
$I = 1000A, V = 415V, \cos ф = 0.8$
∴ P =?
$P = \sqrt{3} x V x I \cos ф$
$= \sqrt{3} x 415 \; x 1000 x 0.8$

$= 575,040.86W \; or \; 575.04kW$

b) *For Apparent Power in VA or kVA*

We know that; $P = \sqrt{3} x V x I$
1000A >18A
$I = 1000A, V = 415V$
∴ P =?
$P = \sqrt{3} x V x I$
$= \sqrt{3} x 415 \; x 1000$

$= 718,801.08VA \; or \; 718.8kVA$

2. *125A*

kW =? *And* kVA =?

a) *For Active power in W or kW*

We know that; $P = \sqrt{3}xVxI \cos \phi$

 125A >27A

 $I = 125A, V = 415V, \cos \phi = 0.8$

 $\therefore P =?$

 $P = \sqrt{3}xVxI \cos \phi$

 $= \sqrt{3}x415 \ x125x0.8$

 $= 71880.1W \ or \ 71.88kW$

b) For Apparent Power in VA or kVA

We know that; $P = \sqrt{3}xVxI$

 125A >18A

 $I = 125A, V = 415V$

 $\therefore P =?$

 $P = \sqrt{3}xVxI$

 $= \sqrt{3}x415 \ x125$

 $= 89850.1VA \ or \ 89.85kVA$

 3.   15A

 $kW =?$ *And* $kVA =?$

a) For Active power in W or kW

 We know that; $P = VxI \cos \phi$

 15A<27A

 $I = 15A, V = 230V, \cos \phi = 0.8$

 $\therefore P =?$

 $P = VxI \cos \phi$

 $= 230 \ x15x0.8$

 $= 2,760W \ or \ 2.76kW$

b) For Apparent Power in VA or kVA
We know that; $P = VxI$
15A<18A
$I = 15A, V = 230V$
$\therefore P = ?$
$P = VxI$
$= 230\,x15$

$= 3,450VA \ or \ 3.45kVA$

Assignment;

Find out the power (P) in terms of kW and kVA
900A, 227A, 75A, 25A.

Thumb Rule for Calculation of Current

Secondary Side / LT Side / Star Side / Output Side of Transformer.

1. Conversion of kW to Amp
- $kW * 1.74 = $ _____Amp (3φ)
- $kW * 5.43 = $ _____Amp (1φ)

2. Conversion of kVA to Amp
- $kVA * 1.39 = $ _____Amp (3φ)
- $kVA * 4.32 = $ _____Amp (1φ)

3. Conversion of kVAR to Amp

- $kVAR * 2.32 = \underline{\hspace{2cm}} Amp \ (3\phi)$

Primary Side / HT Side / Delta Side / Input Side of Transformer

- $kVA * 0.052\underline{\hspace{1cm}} Amp \ (If \ voltage \ (or) POS \ is \ 11kV)$
- $kW * 0.0174\underline{\hspace{1cm}} Amp \ (If \ voltage \ (or) POS \ is \ 33kV)$

Examples

Solve The Following;

Secondary Side / LT Side / Star Side / Output Side of Transformer

1. Conversion of kW to Amp

- $kW * 1.74 = \underline{\hspace{2cm}} Amp \ (3\phi)$

$kW = 2000, I =?$

$\div I = 2000 x 1.74$

$\quad = 3,480A$

- $kW * 5.43 = \underline{\hspace{2cm}} Amp \ (1\phi)$

$kW = 6.5, I =?$

$\div I = 6.5 x 5.43$

$\quad = 30.295A$

2. Conversion of kVA to Amp

- $kVA * 1.39 =$ _____Amp (3φ)

$kVA = 125, I = ?$

∴$I = 125x1.39$
  $= 173.75A$

- $kVA * 4.32 =$ _____Amp (1φ)

$kVA = 8, I = ?$

∴$I = 8x4.32$
  $= 34.56A$

3. Conversion of kVAR to Amp

- $kVAR * 2.32 =$ _____Amp (3φ)

$kVAR = 3, I = ?$

∴$I = 3x2.32$
  $= 6.96A$

Primary Side / HT Side / Delta Side / Input Side of Transformer

- $kVA * 0.052$_____Amp (If my $voltage$ is $11kV$)

$kVA = 1000 , I = ?$

∴$I = 1000x0.052$
  $= 52A$

- $kW * 0.0174$_____Amp (If my $voltage$ is $33kV$)

$kVA = 750 , I = ?$

∴$I = 750x0.0174$
  $= 13.05A$

ELECTRICAL DESIGNING

Introduction to Electrical Designing and Drafting

Electrical: Electrical engineering is a field of engineering that deals with the study of application of electricity and electronics.

Designing: An electrical designing is the design of different electrical systems and components, which are connected to carry out some duties.

Drafting: an electrical drafting, is a type of technical drawing that will provide the outline of lighting and show power layout for an engineering project.

Basically, Electrical is a Module of Designing for rooting or connecting the Electrical and Electronic system throughout your building, industry, plants and factory.

Introduction to Building Industry

(i) Design of a large building is an Extremely complex task, it may take months even years and may include number of Engineers, Supervisors and Technicians.

23

(ii) The designing of residential building is much simpler and may involves a few as one or two Engineers.

(iii) The design of electrical system for large projects is responsibility of electrical consultant company. Consultant people may also carry out other duties such as cost estimation, bill of quantity, bill of materials and field supervision of installation.

(iv) Each of the tasks will be performed in coordination with the Architects or senior MEP Engineer who will carry out over all building planning and designing.

(v) Coordination of work between the Architects and other department is an important and difficult task.

(vi) An error in coordination will give you bad results.

Designing of Electrical System Involves

(i) Survey of area and type of building or structure.

(ii) Total Connected Load (TCL).

(iii) Single Line Diagram (SLD) or One Line Diagram (OLD).

(iv) Number of lighting fixtures required.

(v) Circuit breakers Sizing.

(vi) Transformer Sizing

(vii) Diesel generator sizing.

(viii) UPS sizing.

(ix) Battery sizing.

(x) Cable Sizing.

(xi) Circuit breaker sizing.

(xii) Voltage drop.

(xiii) Short circuit.

(xiv) Tripping time of circuit breaker.

(xv) Load balancing sheet / DB schedule for lighting load and Raw Power Load

(xvi) Capacitor bank sizing.

(xvii) Earthing Calculation.

(xviii) Lighting Arrestor.

(xix) Bus bar Sizing.

(xx) Cable Tray.

Note: Basically, voltage drop calculation and short circuit calculation are usually done if required otherwise then this value will always be within the range.

Requirements to Perform the Electrical Designing

(i) Architectural drawing.

(ii) Civil drawing or Structural drawing.

(iii) Reflected ceiling plan or False ceiling.

(iv) Furniture drawing.

(v) Design brief report.

Introduction of Work Flow of Building Industry / Building Service.

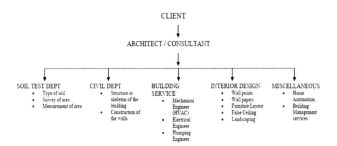

LOAD CALCULATION

Why Should We Need Load Calculations?

1. Sizing of incomer source
   a) Ministry supply
   b) Generator
2. Prepare the load distribution
3. Selection of protection device

How To Do Load Calculation for Office / Buildings

1. Sizing of Incomer Source;

For example;

I need power supply for my office and I am choosing with generator supply. Now I need to find generator sizing?

In this case I have to prepare load calculation first.

Load calculation

Making load calculation we should have load details

No.	Electrical appliances	Watts	Quantity	Total Watts
1	Air conditioner, 2 TON, 1-φ	2400	5	12000
2	Tube lights	72	20	1440
3	Personal Computer (PC)	1600	15	24000
4	Printer	500	3	1500
5	Geezer	1500	3	4500
6	Microwave Oven	1400	1	1400
7	Electric Kettle	2400	1	2400

∴ Total Watts = 47240 Watts
= 47.24 kW

So, we know our total load is 47.24 kW

Required generator size is 125% of total load i.e., 60 kW / 75 kVA generator is needed.

2. Load Distribution;

You have to distribute your load in balanced way otherwise its leads to voltage drop and reduce the efficiency of generator.

For example;

We have 60 kW total load and we have 3-ɸ 415v, 50 Hz supply. Then how do we distribute the load?

Total Load = 60kW

Connecting load	R PHASE	Y PHASE	B PHASE
Load 1 (10 kW)	10		
Load 2 (10 kW)		10	
Load 3 (10 kW)			10
Load 4 (10 kW)	10		
Load 5 (10 kW)		10	
Load 6 (10 kW)			10

3. Selection of Protective Device for Incomer;

If you know the total load and load distribution, then you can easily select the protective device.

There are 5 main protective devices for incomer;

(i)   ELCB
(ii)   RCCB
(iii) MCCB
(iv)   Isolator
(v)   MCB

Based on the load we can select our main incomer as;

ELCB and MCCB

For example:

Total load is 60 kW then ELCB rating should be:

4 pole 200 Amp

Note: Using power formula; P = 1.732 x VI Cos φ

How to Find Amperes

How to convert: kW → A

kVA → A

HP → A

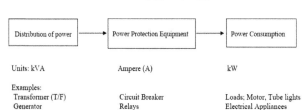

| Distribution of power | Power Protection Equipment | Power Consumption |

Units: kVA   Ampere (A)   kW

Examples:
Transformer (T/F)   Circuit Breaker   Loads; Motor, Tube lights
Generator   Relays   Electrical Appliances

(i)   For 3- φ Supply

	1 kW to Amp	1 kVA to Amp	1 HP to Amp
Formula	$P / \sqrt{3} \times V \times P.f \times E.f$	$P / \sqrt{3} \times V$	$P / \sqrt{3} \times V \times P.f \times E.f$
	$1000 / \sqrt{3} \times 433 \times 0.8 \times 0.9$ =1.851 Amp ≈ 1.9 or 2 A	$1000 / \sqrt{3} \times 433$ =1.33 Amp ≈ 1.4 A	$746 / \sqrt{3} \times 433 \times 0.8 \times 0.9$ =1.38 Amp ≈ 1.5 A

NB: Power Factor (P.f) = 0.8

Efficiency (E.f) = 0.9

(ii)   For 1- φ supply

	1 kW to Amp	1 kVA to Amp	1 HP to Amp
Formula	P / V x p.f 1000 / 230 x 0.6 =7.246 A ≈ 8 A	P / V 1000 / 230 =4.3 A ≈ 6 A	P / V x P.f x E.f 746 / 230 x 0.6 x0.9 =6 A ≈ 6 A

NB: Power Factor (P.f) = 0.6

Efficiency (E.f)= 0.9

Wire Size Calculation and Circuit Breaker Selection

Two Bedroomed House and Kitchen Layout

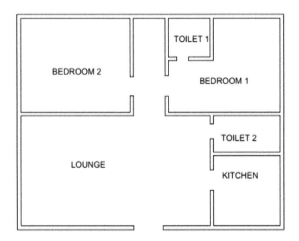

Types of Loads in The House

(i)   Lighting loads (L.L): LED, fans, ex-fans, tubes etc.

(ii)   Power loads (P.L): Sockets; Iron, geysers, chargers, trimmers etc.

(iii)   A/C loads: A/C

For the two bedroomed house and kitchen:

1. Bedroom One (BR_1);

(i)   L.L;   15 W LED x 4 No.   = 60 W

    5 W LED x 2 No.   = 10 W

    60 W fan x 1 No.   = 60 W

T_1 = 15 W LED x 1 No.   = 15 W

    60 W EX-fan x 1 No. =60 W

∴ Total BR_1 L.L is;   =205 W

    =0.205 kW

Because we are using a 3 ф supply, then to A;

= 0.205 x 8 =2 A

6 A SP MCB is recommended.

From the table: 2.5 mm^2 – Phase (ф)

1.5 mm^2 – Neutral

2.5 mm^2 – Earth

BR_1 + T_1 = L.L; 6 A MCB (2.5 sq mm phase wire)

(ii)   (a) BR_1 $P.L_1$

    Iron   = 500W

Hair drier   = 350W

Laptop   = 200W

∴ Total $P.L_1$ = 1050W

    =1.050kW

To find A = 1.050 X 8 = 9 A

⁂ 16 A MCB S.P is recommended

From the table; 4 sq mm^2 – Phase (φ)

2.5 sq mm^2 – Neutral

4 sq mm^2 – Earth

(b) BR$_1$ Toilet P.L$_2$

Geysers =1000 W

Trimmer = 200 W

⁂   Total P.L$_2$ =1200 W

=1.2 KW

To find A, 1.2 x 8 = 9.6 A

⁂ 16 A MCB S.P is recommended

From the table; 4 sq mm^2 – Phase (φ)

2.5 sq mm^2 – Neutral

4 sq mm^2 – Earth

P.L; BR$_1$ (P.L$_1$) = 16 A MCB SP

T$_1$ (P.L$_2$) = 16 A MCB SP

(iii)   AC load

BR$_1$ = 1.4 kW

To find Amperes: 1.4 x 8 = 11.2 A

⁂ 25 A MCB SP is RMD.

From the table; 6 sq mm^2 – Phase (φ)

4 sq mm^2 – Neutral

6 sq mm^2 – Earth

∴ BR₁ AC; 25 A MCB SP

We have two types of loads:

(i)   Inductive load

For 1 kW → A

P.f @ 0.6

P = VI Cos ɸ

I = P / V Cos ɸ

= 100 / 230 x 0.6

= 8A - 3 ɸ

(ii)   Resistive load

1 kW → A

P.f @ 0.8

P= VI Cos ɸ

I = P / V Cos ɸ

=1000/230x 0.8

= 6A -1ɸ

Selection Table

MCB (Amp)	Wires sizes (Sq mm)		Earthing
	Phase	Neutral	
6	2.5	1.5	
10	2.5	1.5	Same
16	4	2.5	As
20	4	2.5	Phase
25	6	4	wire
30	6	4	Size
40	10	6	
50	10	6	
63	16	8	

ELECTRICAL DESIGNING OF HIGH-RISE BUILDING

(a). Five Storey Building, and Ground Floor

(b). Ten Storey Building, and Ground Floor

There are 5 Steps Involved

(i)   Total connected load.

(ii)   Transformer sizing.

(iii)   Circuit breaker sizing of transformer

(iv)   Circuit breaker sizing of each load

(v)   Single line diagram

(a). **Five Storey Building (G +5),** G → Ground floor

+5 → Floors

It contains 2 BHK, 3 BHK, 4 BHK flats

For 2BHK → 1- ϕ Supply

3 BHK & 4 BHK → 3 - ϕ Supply

Energy meter (E.M)

Energy meter is provided by government.

Standards of energy meter (India/Kenya/Gulf)

Flats	Villas
2 BHK – 5 KW	3 BHK – 10MKW
3 BHK – 7.5 KW	4 BHK – 15 KW
4 BHK – 10 KW	

Example

Five storey building buildings

Three 3 BHK flats on each floor, 15 flats

Three 4 BHK flats on each floor, 15 flats

(i)   Total connected load (TCL)

TCL = Total flat load + Total common load.

(a) Total common load

Lift   = 7.5 H.P × 2 No.
Bore hole pump = 5 H.P × 1 No.
Water pump   =5 H.P × 1 No.
Lighting   = 6 kW

∴   Total common load,   Lift   = 7.5 × 0.746 KW × 2 =12 KW

B. Pump $= 5 \times 0.746$ KW $\times 1 = 4$ KW

W. Pump $= 5 \times 0.746$ KW $\times 1 = 4$ KW

Lighting $= 200$ (No. of lighting points) $\times 30$ W $= 6$ KW

∴ T.C.L $= 26$ KW

(b) Total load of flats

3 BHK $= 7.5$ kW $\times 15 = 112.5$ kW ≈ 113 kW

4 BHK $= 10$ kW $\times 15 = 150$ kW

∴ T.L.F $= 113$ kW $+ 150$ kW $= 263$ kW

∴ T.C.L $= 26$ kW $+ 263$ kW

$= 289$ kW

(c) Potential Demand / Peak Demand (P.D)

Demand factor $= 60$ % (recommended)

Demand factor $=$ peak demand (P.D) / T.C.L

∴P.D $=$ demand factor $\times$ T.C.L

$= 0.6 \times 289$

$= 173.4$ kW

≈ 174 kW

Demand factor (D.f)

	India	Gulf
Residential	50% to 70%	90%
Commercial	70% to 90%	90%
Industry	125%	125%

(ii)   Transformer sizing

P.D = 174 KW

kVA = kW/ P. f = 174 /0.8 = 218 kVA

Thus, from the table; 315 KVA T/f is recommended.

(iii)   Circuit breaker (CB) sizing of Transformer (T/f)

315 KVA × 1.4 A = 441 A

≈ 500 A (MCCB, T.P) for meter panel

	Multiplying Factor	
	3φ	1φ
KVA to A	1.4	6
HP to A	1.5	6
KW to A	1.9	8

(iv) Circuit breaker sizing for load

3 BHK, 7.5 kW

$= 7.5 \times 1.9 \times 2$ {Safety multiplying factor}

$= 29A$

From the standard chart;

50 A MCB 4P is recommended.

4 BHK, 10 kW

$= 10 \times 1.9 \times 2$ {Safety multiplying factor}

$= 38A$

From the standard chart;

63A MCB 4P is recommended.

Common load – 26 KW

$= 26 \times 1.9 \times 2$ {Safety multiplying factor}

$= 99A$

From the table,

125 A MCCB is recommended.

(v) Single Line Diagram (S.L.D / O.L.D)

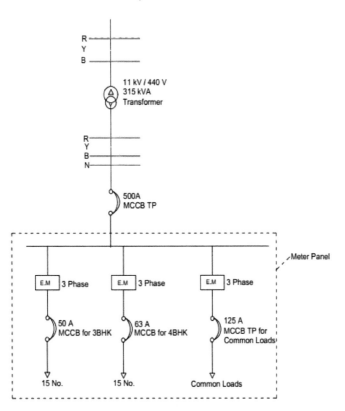

(b). Ten Storey Building (G + 10), G → Ground Floor

$$+10 → \text{Floors}$$

Steps:

(i)   Total   connected   load (T.C.L)

T.C.L = Common + Flat loads

(a) Common load

Lift   = 10 HP x 3 No.   = 10 x 0.746 kW x 3 = 22.38 kW

Borehole pump   = 5 HP x 2No.   = 5 x 0.746 kW x 2 = 7.46 kW

Water pump   = 5 HP x 4 No.   = 5 x 0.746 kW x 2 = 14.93 kW

Lighting parking   = 20 kW   =   20 kW = 20 kW

∴ T. Common load = 64.76 kW   ≈65kW

(b) Flat load (total)

4 x 2 BHK each floor,   5 kW x 4 x 10 = 200 kW

3 x 3 BHK each floor,   7.5 kW x 3 x 10 = 225 kW

3 x 4 BHK each floor,   10 kW x 3 x 10 = 300 kW

∴ Total connected load = common load + Flat load
= (65 kW + 200 kW + 225 kW + 300 kW)
= 790 kW

43

(c) Peak demand / Potential demand (P.D)

D.f = Peak demand or Potential demand / T.C.L

∴ Peak demand = D.f x T.C.L , D.f =70%
  =0.7
  = 0.7 x 790 kW
  = 553 kW

(ii) Transformer sizing

P.D = 553 kW

kVA = kW / P.f = 553 kW / 0.8 = 691.25 kVA ≈ 692 kVA

∴ From the standard chart; 800 kVA T/F is recommended.

(iii) Circuit breaker sizing T/F

T/f, 800 kVA

∴ 800 kVA x 1.4 = 1120 A

From the table; 1250 A ACB T.f is recommended.

(iv)   (a) Circuit breaker sizing for main meter panel flats

10 Storey Building and Ground, Sketch Diagram

10	Panel 4
9	
8	
7	Panel 3
6	
5	
4	Panel 2
3	
2	
1	Panel 1
C.L	

Panel 1

1st floor + C.L, =10 flats + CL

⸪10 flats;   2 BHK x 4 = 5 kW x 4   = 20 kW

3 BHK x 3 = 7.5 kW x 3   = 22.5 kW

4 BHK x 3 = 10 kW x 3   = 30 kW

C.L   = 60 kW   = 60 kW

⸫ Total value   = 137.5 kW ≈ 138 kW

D.f   = 70%

T.C.L = 138 kW

⸫ P.D   = D.f x T.C.L

= 0.7 x 138 kW

= 97 kW

⁂ C.B for panel 1 = P.D x 1.9

$$= (97 \times 1.9) \text{ A}$$

$$= 184.3 \text{ A} \approx 185 \text{ A}$$

⁂ 250 A triple pole MCCB is recommended.

For panel 2,3 & 4 (equal)

Each floor 10 flats, =10 EM

$$2BHK \times 4 = 5kW \times 4 \times 3 = 60 \text{ kW}$$

$$3 \text{ BHK} \times 3 = 7.5kW \times 3 \times 3 = 67.5 \text{ kW}$$

$$4 \text{ BHK} \times 3 = 10 \text{ kW} \times 3 \times 3 = 90 \text{ kW}$$

⁂ Total load » 217.5 kW ≈ 218 kW

D.f = 70%

P.D   = D.f x T.C.L

$$= 0.7 \times 218 \text{ kW}$$

$$= 152.6 \text{ kW} \approx 153kW$$

⁂ C.B for each panel (2,3,4) = 153 kW x 1.9

$$= 291 \text{ A}$$

⁂ 400 A MCCB T/P is recommended.

(b) Circuit breaker sizing for each load

2 BHK = 5 KW x 1.9 x 2 (safety factor)

= 19A

∴ 40 A four pole MCB is recommended.

3 BHK = 7.5 KW x 1.9 x 2 (safety factor)

= 28.5 A

∴ 50 A four pole MCB is recommended

4 BHK = 10 KW x 1.9 x 2 (safety factor)

= 38A

∴ 63 A four pole MCB is recommended

C.L = 65 KW x 1.9 x 2 x 0.7 {D.f}

= 173 A

∴ 200 A MCCB TP is recommended.

Single line diagram / One line diagram (SLD/ OLD)

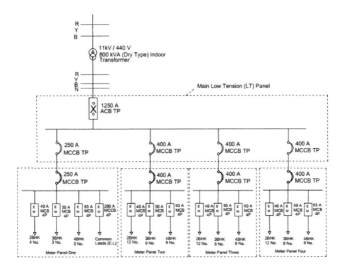

Panel 1

Total energy meters;

$$(4 \text{ flats x } 1) + (3 \text{ flats x } 1) + (3 \text{ flats x } 1) + \text{C.L}$$

$$= 11 \text{ E.M}$$

Panel 2, 3 & 4

Total energy meters;

(4 flats x 3) + (3 flats x 3) + (3 flats x 3)

$$=30 \text{ E.M}$$

N.B: A max of 30 E.M is required for each panel.

APFC PANEL DESIGNING

APFC Panel kVAR Calculation

$P.F = \cos \phi = $ Active power/ Reactive power

System 100% power; its P.F is 0.85

Then the net power in the system is 85% active power, 15 % reactive power.

Two Methods

There are two methods of kVAR calculation;

 (i) Classic calculation method
 (ii) Simple table method

 (i) **Classic calculation method**

Required $kVAR = P(\tan \phi_1 - \tan \phi_2), \quad (\tan \phi_1 - \tan \phi_2)$ is a Multiplying factor

$C = Farad, C = (VAR/2\pi f v^2)$

$F = Frequency$

$P = Real\ Power\ in\ kW$

$V = Voltage\ in\ Volts$

Example

$Load = 100kVA$, $P.f = 0.8 \ (Measured)$, $P.f = 0.9 \ (Required)$

$\cos \phi_1 = Measured \ P.f$, $\cos \phi_1 = 0.8$, ∴ $\phi_1 = \cos^{-1} 0.8 = 36.86^o$

$\cos \phi_2 = Required \ P.f$, $\cos \phi_2 = 0.9$, ∴ $\phi_2 = \cos^{-1} 0.9 = 25.84^o$

$P = Power \ in \ kW$

$kVA = \frac{kW}{Pf}$

∴ $kW = kVA * Pf$

∴ $kW = 100x0.8$

$= 80kW$

∴ $kVAR \ \text{Required} = P(\tan \phi_1 + \tan \phi_1)$

$= 80(\tan 36.86 - \tan 25.84)$

$= 80(0.257)_{MF}$

$= 20.56 \ kVAR$

(ii)   **By Simple table method**

Standard Table for Capacitor Value Calculation

	0.9	0.91	0.92	0.93	0.94	0.95	0.96	0.97	0.98	0.99	1
0.6	0.849	0.878	0.907	0.938	0.97	1.005	1.042	1.083	1.13	1.191	1.333
0.61	0.815	0.843	0.873	0.904	0.936	0.97	1.007	1.048	1.096	1.157	1.299
0.62	0.781	0.81	0.839	0.87	0.903	0.937	0.974	1.015	1.062	1.123	1.265
0.63	0.748	0.777	0.807	0.837	0.87	0.904	0.941	0.982	1.03	1.09	1.233
0.64	0.716	0.745	0.775	0.805	0.838	0.872	0.909	0.95	0.998	1.058	1.201
0.65	0.685	0.714	0.743	0.774	0.806	0.84	0.877	0.919	0.966	1.027	1.169
0.66	0.654	0.683	0.712	0.743	0.775	0.81	0.847	0.888	0.935	0.996	1.138
0.67	0.624	0.652	0.682	0.713	0.745	0.779	0.816	0.857	0.905	0.966	1.108
0.68	0.594	0.623	0.652	0.683	0.715	0.75	0.787	0.828	0.875	0.936	1.078
0.69	0.565	0.593	0.623	0.654	0.686	0.72	0.757	0.798	0.846	0.907	1.049
0.7	0.536	0.565	0.594	0.625	0.657	0.692	0.729	0.77	0.817	0.878	1.02
0.71	0.508	0.536	0.566	0.597	0.629	0.663	0.7	0.741	0.789	0.849	0.992
0.72	0.48	0.508	0.538	0.569	0.601	0.635	0.672	0.713	0.761	0.821	0.964
0.73	0.452	0.481	0.51	0.541	0.573	0.608	0.645	0.686	0.733	0.794	0.936
0.74	0.425	0.453	0.483	0.514	0.546	0.58	0.617	0.658	0.706	0.766	0.909
0.75	0.398	0.426	0.456	0.487	0.519	0.553	0.59	0.631	0.679	0.739	0.882
0.76	0.371	0.4	0.429	0.46	0.492	0.526	0.563	0.605	0.652	0.713	0.855
0.77	0.344	0.373	0.403	0.433	0.466	0.5	0.537	0.578	0.626	0.686	0.829
0.78	0.318	0.347	0.376	0.407	0.439	0.474	0.511	0.552	0.599	0.66	0.802
0.79	0.292	0.32	0.35	0.381	0.413	0.447	0.484	0.525	0.573	0.634	0.776
0.8	0.266	0.294	0.324	0.355	0.387	0.421	0.458	0.499	0.547	0.608	0.75
0.81	0.24	0.268	0.298	0.329	0.361	0.395	0.432	0.473	0.521	0.581	0.724
0.82	0.214	0.242	0.272	0.303	0.335	0.369	0.406	0.447	0.495	0.556	0.698
0.83	0.188	0.216	0.246	0.277	0.309	0.343	0.38	0.421	0.469	0.53	0.672
0.84	0.162	0.19	0.22	0.251	0.283	0.317	0.354	0.395	0.443	0.503	0.646
0.85	0.135	0.164	0.194	0.225	0.257	0.291	0.328	0.369	0.417	0.477	0.62
0.86	0.109	0.138	0.167	0.198	0.23	0.265	0.302	0.343	0.39	0.451	0.593
0.87	0.082	0.111	0.141	0.172	0.204	0.238	0.275	0.316	0.364	0.424	0.567
0.88	0.055	0.084	0.114	0.145	0.177	0.211	0.248	0.289	0.337	0.397	0.54
0.89	0.028	0.057	0.086	0.117	0.149	0.184	0.221	0.262	0.309	0.37	0.512
0.9		0.029	0.058	0.089	0.121	0.156	0.193	0.234	0.281	0.342	0.484
0.91			0.03	0.06	0.093	0.127	0.164	0.205	0.253	0.313	0.456
0.92				0.031	0.063	0.097	0.134	0.175	0.223	0.284	0.426
0.93					0.032	0.067	0.104	0.145	0.192	0.253	0.395
0.94						0.034	0.071	0.112	0.16	0.22	0.363
0.95							0.037	0.078	0.126	0.186	0.32

Required $kVAR = kW \; x \; Table \; Multiplier \; Value$

$kW = 80$

From table; Power factor from 0.8 – 0.9, Multiplying factor is found to be; 0.266

∴$kVAR$ Required $= 80(0.266)_{M.F}$

$$= 21.26 kVAR$$

Example

APFC Panel

2500 kVA T/f, 33kV/440V

Thumb rule:

40% of 2500 kVA = 1000 kVAR ←Capacitor bank value

30% of 2500 kVA =750 kVAR {Variable capacitor band} – Load

10 5 of 2500 kVA =250 kVAR {Fixed capacitor Bank} – No load

Standard size of capacitors

I5 kVAR	30kVAR
10 kVAR	40 kVAR
15 kVAR	50 kVAR
20 kVAR	100 kVAR
25 kVAR	**{Very less available}**

Steps to Design APFC Panel

(i)   Find the value of capacitor

i.e., 1250 kVA T/f

40% of 1250 kVA = 500 kVAR – Value of capacitor Bank

30% of 1250 kVA = 375 kVAR – Variable capacitor bank-Load

10% of 1250 kVA = 125 kVAR – Fixed capacitor bank- no load

(ii)   Find the value of circuit breaker of T/f

1250 kVA X 1.4 = 1750 A   NB: 3ф,

kVA $\rightarrow$ A$\rightarrow$ 1.4_{MF}

2000A ACB is recommended   HP

$\rightarrow$ **A** $\rightarrow$ 1.5_{MF}

                         kW $\rightarrow$

A $\rightarrow$ 1.9_{MF}

(iii)   Find the value of APFC panel C.B

500 kVAR $\rightarrow$ Total capacitor value   P= $\sqrt{3}$VISin 90^0

C.B for APFC panel = kVAR $_T$ x 2_{MF} I= P/$\sqrt{3}$VSin 90^0

 =500X 2   = 1000/$\sqrt{3}$
x 440x sin 90^0

=1000A ACB   =1.31A

≈2A

NB: P=1kW

(iv)   Single line diagram

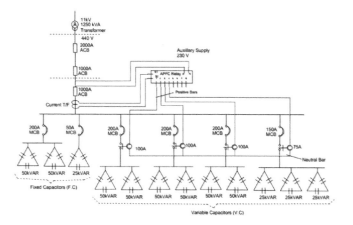

NB:

- C → CDC → Capacitor duty conductor
- Fixed capacitors use Circuit Breakers (C.B) only.
- Variable capacitors use C.B, CDC and APFC Relay.

Standard Tables/Charts

Transformer standard sizes

1	63 Kva	13	950/100 kVA
2	80 kVA	14	1250 kVA
3	100 kVA	15	1500 kVA
4	125 kVA	16	1800 kVA
5	160 kVA	17	2000 kVA
6	200 kVA	18	2500 kVA
7	250 kVA	19	3150 kVA
8	315 kVA	20	4000 kVA
9	400 kVA	21	5000 kVA
10	500 kVA	22	6300 kVA
11	630 kVA	23	7500 kVA
12	750/800 kVA	24	1000 kVA

Circuit breakers sizes

MCB (A)	MCCB (A)	ACB (A)
0.5	8	630
2	16	800
4	32	1000
6	40	1250
10	80	1600
15/16	100	2000
20	125	2500
25	160	3200
30/32	200	4000
40	250	5000
50	320	6300
60/63	400	
	500	
	630	
	800	
	1000	
	1250	

Lux and Lumen Calculation in Details

How to Calculate Lux and Lumen

Lux: - The lux (l) is the unit of luminance in the international system of units (SI)

Lumen (LM): - It is the SI unit of luminous flux.

Luminous flux: - The amount of light energy or brightness that can be emitted from bulbs or lamp is known as luminous flux (lumen).

Formula: - $= \dfrac{Length\ (l)\ x\ Width\ (w)\ x\ Lux\ level}{Cuf\ x\ M.f\ x\ Total\ lumens\ x\ Type\ of\ Fixtures}$

Recommended Lux level:

Building space	Lux level
Bedroom	240
Church	240
Car parking	40
Laboratory	400

Cuf: - Coefficient of utilization factor

The amount of light energy that is falling on working place is called coefficient of utilization.

Room index: - Ratio of L, W and HM.

HM: - Mounting light of lamp/ fixture.

Lumen calculation

I.R = L x W / (L+W) HM

Fixture mount type: -

Fixture Mount Type		Wall Color Type	
Ceiling	70%	Very light	70%
Wall	50%	Normal light	50%
Foot / Floor	20%	Dark	20%

Maintenance factor: -

M.f is based on environment and lamp, L.L.f (light lose factor)

Bedroom- Neat –   5%

Kitchen – Normal -   20%   **Formula;** 100% - Value e.g., for bedroom

Factory – Untidy-   30%
100%- 5% = 95%

Total Lumen: -

Types of lamps	Lm for Watt
1. Tungsten filament lamp	20 Lm/W
2. C.F.L	60 Lm/W
3. LED	80 Lm/W

Total lumen: - For 15 W / 12 W / 18 W LED

15 W x 80 Lm/W = 1200 Lumen

12 W x 80 Lm/W = 960 Lumen

18 W x 80 Lm/W = 1440 Lumen

Number of fixtures: -

The amount of light that should be installed in an area is known as number of fixtures.

Type of fixture: -

Number of lights in a unit.

Example

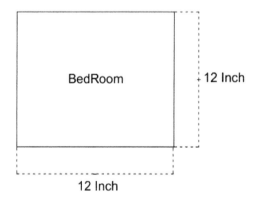

To convert feet into Meters;

12' x 0.304 = 3.65M;

L= 3.65 M, W= 3.65 M, Lux level = 240 Lm

Using 12 W LED

∴T.L = 12 W x 80 Lm/W = 960 Lumen

Number of fixtures $= \dfrac{L \; x \; W \; x \; Lux \; level}{Cuf \; x \; M.f \; x \; T.L \; x \; Type \; of \; Fixture}$

Cuf: - Room index = L x W / (Lx W) HM

$\qquad\qquad$ = 3.65 x 3.65 / (3.65 x 3.65) 2.88

$\qquad\qquad$ i.e., HM =

HM = 12' - 2.5'

$\qquad$ = 9.5'

$\qquad$ Into Meters; 9.5' x 0.304 = 2.88M

Example

0.5 = Wall color type

$\qquad$ Normal light 50%

0.7 = Mount type

$\qquad$ Ceiling 70%

0.65 = Room index

∴ $\text{Cuf} = \dfrac{Mount\ Type + Wall\ Color\ Type + Room\ Index}{3}$

$\quad = \dfrac{0.7 + 0.5 + 0.62}{3}$

$\quad = 0.62$

M. f; 100% -5% = 95% = 0.95 (Because it is a bedroom)

T.L; 960 lm

Type of fixture; 1

∴ Number of fixtures $= \dfrac{L\ x\ W\ x\ Lux\ level}{Cuf\ x\ M.f\ x\ T.L\ x\ Type\ of\ Fixture}$

$\quad = \dfrac{3.65\ x\ 3.65\ x\ 240}{0.62\ x\ 0.95\ x\ 960\ x\ 1}$

$\quad = 5.65 \approx 6$

∴ 12 W LED, 6 numbers is recommended.

How to place the fixtures: -

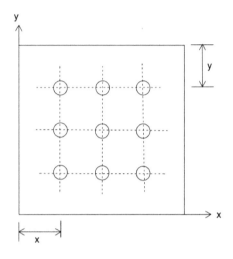

x= (n+1) = (2+1) =3

   3.65M/ 3 = 1.21M or 12'/3 = 4'

y= (n+1) = (3+1) =4

   3.65M/4 = 0.91M or 12'/4 = 3'

Lux Level Standards: -

Circulation Area:

Corridor, Passageway	100
Lift	150
Stair Case	150
Escalator	150

Entrances:

Entrance Halls, Lobbies, Waiting Room	300
Enquiry Desk	500
Gate House	300
Kitchens	300
Food Stores	150
General	300

Outer Door:

Control Entrance Hall or Exit Gate	150
Entrance & Exit Car Parking	30
Stores, Stock Yards	30
Industrial Covered Ways	50

Medical and First Aids Centre:

Consultant Room, Treatment Area	500
Medical Store	100
Rest Room	150

Store & Stock Rooms:

Server Room	300
Pump Room	300
Parking Area	100
Repair Servicing, Greasing, Pits, Washing, Polishing	500

Inspection & Testing Shop:

Rough work	300
Medium work	500
Fine work	1000
Very fine work	1500
Minute work	3000
Laboratories (General)	750

Staff Room:

Changing Room, Lockers Room & Cleaner's Room	150
Rest Room	300

Laundries & Dry-Cleaning Works:

Receiving, Sorting, Washing, Drying, Ironing, Dispatching, Dry Cleaning, Pressing, Inspection, Hand Ironing	300

Pharmaceutical & Fine Chemical Works Pharmaceutical Manufacture:

Grinding, Mixing, Filling, Labeling, Capping, Wrapping & Cartooning	500
Inspection	750
Fine Chemical Manufacture	300
Fine Chemical Finishing	500
Raw Material Store	300

Printing Machine Room:

Presses	500
Premade Ready	500
Printed Sheet Inspection	1000

Rubber Processing Factories:

Preparation Needs, Dipping, Molding	300
Tyro & Tube Making	500

Sheath Metal Works:

Bench work, Scribing, Inspection	750
Pressing, Punching, Shearing, Stamping, Spinning, Folding	500

Welding & Sorting Shop:

Gas & Arc Welding, Spot Welding	300
Medium Soldering, Brazing	500
Fine Soldering	1000
Very Fine Soldering	1500

Wood Working Shop:

Rough Sawing, Bench Work Sizing, Rough Sanding	300
Medium & Bench Work Gluing	500
Fine Machine Work	750

Office:

General Office	500
Deep Plan & General Office	750
Business Machine & Typing	750
Fitting Room	300
Conference Room	750

Slaughter House:

General	500
Inspection	750

Office & Shops:

Executive Office	500
Computer Room	500
Punch Card Room	750
Drawing Offices Drawing Boards	750
Reference Table & General	500
Print Room	300

General Shops & Marts:

Conventional with Counters	500
Self Service	500

Load Details of Difference Appliances Used in Day-to-Day Life, In Watts:

Types of Appliances	Rating in Watts
Washing Machine	400 - 2000
Water Heater	800 – 3200
Micro Wave Oven	800 – 3200
Iron Box	400 – 1600
Dish Washers	400 – 1600
Freezer or Refrigerator	600 – 1200
Mixer or Grinder	400 – 1200
Rice Cooker	400 – 3200
Vacuum Cleaner	600 – 1400
Ceiling Fan	40 – 100
Water Filter	100 – 600
Exhaust Fan	60 – 250
Chimney Kitchen	400 – 1600
Air Cooler	250 – 750
Music System	200 – 700
TV or Monitor	80 – 160
LCD	80 – 120
LED	40 – 100

CABLE TRAY SIZE CALCULATION/ CABLE TRAY SELECTION

We have four main steps;

(i)   Determine cable diameter
(ii) Air space between conductor and cable layout to maintain air spaces.
(iii) Determine width of tray.
(iv) Thermal expansion and contraction.

(i)   Cable Diameter

Formulas

(a) Area of circle $A = \pi r^2$
(b) Diameter of circle $D = 2r$

Example:
300 sq mm x 3.5C x 4R

3 Core
0.5 Neutral

4 - Run

Cable Tray

(a) $A = \pi r^2$, A= 300 sq mm

$$r^2 = {}^A\!/_\pi$$

∴ r= $\sqrt{{}^A\!/_\pi}$ = √300 / 3.14 =9.77 mm

(b) D = 2r
=2 x 9.77
=19.549 mm
Because it's a three core;

∴ 19.549 x 3.5 = 68.42 mm

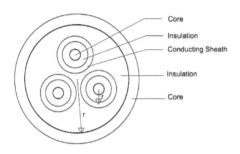

∴ Diameter of the cable; = 68.42 mm

(ii)   Air Space and Cable Layout
Not <25%, Not >100%

68.42 mm 3.5C 4 Run

Spacing = X

X between A <>A; 68.42 x 0.25 = 17.05

(iii) Determine Width of Tray

4 conductors, i.e. A; 68.42 mm X 4 = 273.68 mm

X; 7.05 x 3 = 51.5 mm

∴ Total width of tray = A + X

= (273.68 + 51.5) mm

= 325.18 mm

Cable Tray Standard Sizes

Width	Height
50 mm	50 mm
100 mm	100 mm
150 mm	150 mm
200 mm	200 mm
250 mm	
300 mm	
400 mm	
500 mm	
600 mm	

∴ The cable tray we are recommending is 350x 100 mm cable tray.

(iv) Thermal expansion and contraction

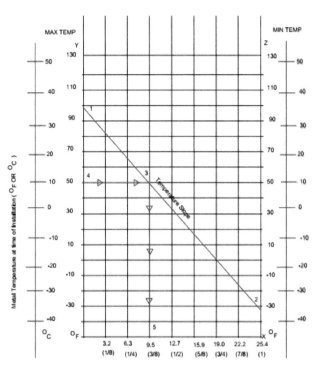

Gap Setting in mm /inches

Gap setting of expansion connector plates 25.4 mm (1^0) Gap Max.

Y- Axis – Maximum temperature

X- Axis – Minimum temperature

In this case; 9.5mm is the gap setting between two cable trays

Expansion couples:

1. **Concerting expansion coupler**
2. **Sliding expansion coupler**

Types of cable trays

(i)   Ladder type -   Powder coated

                           -   Hot dip Galvanized Iron (GI)

(ii)   Mesh type -Normal

   -Hot dip GI

(iii) Perforated type

Indoors   – powder coated

Outdoors – hot dip Galvanized Iron (GI)

Busbar Size and Price Calculation / Busbar Size Chart and Price

(i)   Busbar (BB) calculation

Standard sizes of busbar

Width (mm)	Thickness in (mm)
20	3
25	6
30	10
35	12
40	
50	
75	
100	
120	
150	
180	
200	

Advantages

- Multiple connections
- Compact design
- Reduce cable trays
- Decrease size of main Switch Board (SB)
- Simplified design

Current Carrying Capacity (CCC)

	Aluminium	Copper
Normal	0.8	1.2
Pure Electron	1	1.6

BB = Width x Thickness x CCC

i.e. 63 A to find the correct busbar size;

» W x T x CCC

For Al;

   (a) 20 x 3 x 0.8 = 48 A (Not valid size)
   (b) 30 x 6 x 0.8 = 72 A (valid size)
   (c) 25 x 3 x 0.8 = 60 A (Not valid size)

⁂ For 63 A the correct size of busbar is; 30 x 60 x 0.8 Al

For Cu;

W x T x CCC
        (a) 20 x 3 x 1.2 = 72 A (Valid size)
For 100 A;
30 x 3 x 1.2 = 108 A (Valid size)
For Al; 200 A
200 x 3 x 0.8 = 1920 A (Not valid)

So, for higher currents we multiply the width by 2; i.e., 2000 A

120 x 12 x 0.8 = 1200 A

1200 x 2 = 2400 A ∴ Valid

∴ Two busbars are sandwiched.

(ii)   Kg/M and pricing

BB = W x T x W.D(Width Density)

Aluminium	Copper
0.0027	0.0089

For 100 A, Al

BB = 50 x 3 x 0.8 = 120 ∴ Valid

Kg/M = 50 x 3 x 0.0027 = 0.405 Kg/M

For 10 Metre;

0.405 x 10 = 4.05 Kg/10M

The price of Al per Kg = 400 Rs/Kg

For 4.05 Kg /10 M = 4.05 x 400 = 1620 /10 M

=1620 Rs / 10M

For 100A, Cu

BB; 30 x 3 x 1.2 = 108 A (Valid)

Kg/M = 30 x 3 x 0.0089 = 0.801 Kg

The price of Cu per Kg; 1200 Rs/Kg

For 10 M = 0.801 x 10 = 8.01Kg/10 M

8.01 kg/10 M x 1200 Rs/kg = 9,612 Rs / 10Mtrs

AMMETER AND VOLTMETER WIRING

The required instruments are:

i)   Instrument Transformer (T/f)

-Current T/f ⎤ Measuring Voltage and Current

- Potential T/f ⎦

C.T – step downs current by maintaining constant voltage.

I. Ammeter Circuit diagram

There are 3 components required;

(1) Current Transformer (C.T)
(2) Ammeter Selector Switch (A.S.S)
(3) Digital Ammeter

Circuit diagram

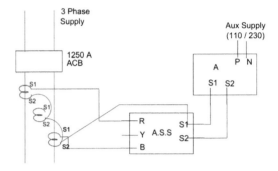

NB:

- C.T changes depending on load
- A.S.S and Ammeter will be or remains the same

II. Voltmeter circuit wiring diagram

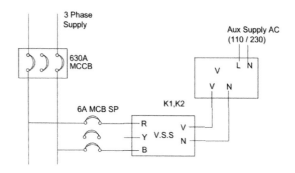

There are 3 components required;

(1) V.S.S – Voltage Selector Switch
(2) Digital Voltmeter (D.V)
(3) Circuit breaker for protection

How To Calculate Fault Current for Electrical System

(1) Calculation for Single Line Diagram (SLD) system

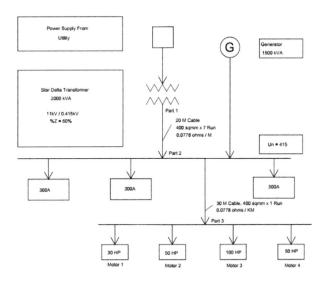

Calculation of short circuit current at point No. 1

Available data

- Line voltage = 415 V
  = 0.415kV
- kVA = 2000
- % Z = 6. 00

Step 1: We will calculate impedance of a transformer in ohms from %Z using the following formula:

$$Z_{Transformer} = \frac{\%Z * 10 * kV^2}{kVA}$$

$$= \frac{6.00 * 10 * (0.415)^2}{2000}$$

$$= 0.005166\Omega$$

Step 2: Calculate short circuit current at point number 1 using following formula:

$$I_{S/C} = \frac{1.05 * Line\ Voltage}{\sqrt{3} * Z_{Transformer}}$$

$$= \frac{1.05 * 0.415}{\sqrt{3} * 0.005166}$$

$$= 48.7kA$$

∗ Fault level at point 1 = 48.7 kA

Calculation of short circuit current at point No. 2

Available data

$Z_{TR} = 0.005166$

Z_{Cable} (Between point number 1 to number 2) = 0.0778 Ω/Km

No. of Runs =7

Distance point 1 to 2 = 20 M = 0.02 KM

$$Z_{Cable} = \frac{(Ohms/KM * Distace\ in\ KM)}{No.\,of\ Runs}$$

$$= (0.0778 * 0.02)/7$$

$$= 0.00022\Omega$$

$$I_{S/C} = \frac{1.05 * Line\ Voltage}{\sqrt{3} * (Z_{Transformer} + Z_{Cable})}$$

$$= \frac{1.05 * 0.415}{\sqrt{3} * (0.005166 + 00022)}$$

$$= 46.1 kA$$

∗ Fault level at point 2 = 46.1 KA

Calculation of short circuit current at point 3
Available Data;

Z_{Tf} = 0.005166

Z_{Cable} (Between point 1 and 2) = 0.00022 Ω

Z_{Cable} (Between point 2 and 3) = 0.0778 Ω/KM

Number of Runs = 1

Distance of point 2 to 3 = 30 M = 0.03 KM

$$Z_{Cable} = \frac{(Ohms/KM * Distace\ in\ KM)}{No.\ of\ Runs}$$

$$= (0.0778 * 0.03)/1$$

$$= 0.002334\Omega$$

$$I_{S/C} = \frac{1.05 * Line\ Voltage}{\sqrt{3} * (Z_{Transformer} + Z_{Cable\ Point\ 1\ to\ 2} + Z_{Cable\ Point\ 2\ to\ 3})}$$

$$I_{S/C} = \frac{1.05 * 0.415}{\sqrt{3} * (0.005166 + 0.00022 + 0.002334)}$$

$$= 33.5\ kA$$

∴ Fault level at point 3 = 33.5 kA

SINGLE LINE / SCHEMATIC DIAGRAM DESIGN

- What is single line diagram?
- Basic symbols
- How to draw single line diagram

(i)   What is single line diagram?

The process of representation of a system with graphical (with using symbols) is called single line diagram (SLD) or one line diagram (OLD) or schematic diagram

i.e.

For a restaurant complex below;

SMDB - R1 Restaurant - 1	SMDB - R2 Restaurant - 2	Spare Restaurant - 3	Spare Restaurant - 4
SMLV Room	Pump Room Water Tank	Spare Restaurant - 5	

(ii)   Symbols

	Transformer
	Generator
	Multi-step power factor correction capacitor bank
	Isolation transformer
	Air circuit breaker (three phase)
	Molded case circuit breaker (three phase)
	Molded case circuit breaker (single phase)
	Residual current circuit breaker (single phase)
	Residual current device
	Variable frequency device
	Single phase Isolator
	Three phase Isolator
	Final distribution board
	Motor control panel (MCP)
	Sub-main distribution board (SMDB)

(iii)   How to draw single line diagram

Let us consider one example to draw a single line diagram of a restaurant complex.

MDB having the following load/ feeders as:

SMDB- R1 – 300kW

SMDB- R2 – 300kW

SMDB- R-E – 50kW

SPARE   - 227 kW

SPARE   - 227 kW

SPARE   - 227 kW

Capacitor bank size 375 kVAR

Further SMDB-R-E sub load as follows:

DB-S   -2 kW

Jump pump – 1.5 kW

PR-EF-01   – 0.2 kW

PR- MAF -01   - 0.2kW

Booster pump – 5kW

Fire pump – 42.5 kW

Calculation and Sizing

Step-1: Total Connected load

Total load; 300 + 300 + 50 + 227 + 227 + 227 = 1333 kW

Maximum demand   = Total load x demand factor

                     =1333 x 0.9 = 1199.7 kW

                     ≈ 1200 kW

Step-2: Transformer Sizing

As per ADDC and DEWA regulation, we have transformer size and its rated load.

(a) 60 Amp feeder   - 30 kW
(b) 100 Amp feeder   - 50 kW
(c) 125 Amp feeder   - 60 kW
(d) 160 Amp feeder   - 80 kW
(e) 200 Amp feeder   - 100 kW
(f) 300 Amp feeder   - 150 kW
(g) 400 Amp feeder   - 200 kW
(h) 1000 Amp feeder   - 800 kW
(i) 1500 Amp feeder   - 1200 kW

Therefore, transformer size for our case is 1500 kVA

Step-3: Breaker Selection

For load 300kW; P = √3 x V x I x Cos φ

I = 300 x 1000 /√3 x 400 x 0.8

  = 300000/ 554.25

 = 541 A

I_n = 630 A (next breaker size)

Short cut Method

Break size = Total load x 2

        = 300 x 2 = 600 A ≈ 630 A (next breaker size)

Similar we can calculate breaker sizes

SMDB- R1 – 300kW =630 A

SMDB- R2 – 300kW = 630 A

SMDB- R-E – 50 kW = 100 A

SPARE   - 227 kW = 630 A (set as 500 A)

SPARE   - 227 kW = 630 A (set as 500 A)

SPARE   - 227 kW = 630 A (set as 500 A)

Step-4: Cable Selection

Breaker (A)	CS-4C	CS-1C	CS-
12	4C x 4	2C x 4	1C x 4
16	4C x 4	2C x 4	1C x 4
20	4C x 4	2C x 4	1C x 4
32	4C x 6	2C x 6	1C x 6
40	4C x 10	2C x 10	1C x 10
63	4C x 16	2C x 16	1C x 10
80	4C x 25	2C x 25	1C x 16
100	4C x 35	2C x 35	1C x 16
125	4C x 50		1C x 25
150	4C x 70		1C x 35
160	4C x 70		1C x 35
200	4C x 95		1C x 50
225	4C x 120		1C x 70
250	4C x 150		1C x 70
300	4C x 185		1C x 95
350	4C x 240		1C x 120
400	4C x 300		1C x 150

Our load current for 300kW load is 541A. From the chart cable size is; 185mm^2

We need to run 2 cables 185mm^2 with earth cable 1C x 95mm^2

Step-5: Capacitor Bank Calculation

Capacitor bank size = Total load {tan (Cos^{-1} (old power factor))–tan (Cos^{-1} (new power factor))}

Load = 1333 kW, old power factor = 0.8, new power factor = 0.9

$$= 1333 \text{ x } \{\tan (\text{Cos}^{-1} (0.8)) -\tan (\text{Cos}^{-1} (0.9))\}$$

$$\approx 1333 \text{ x } (0.74 - 0.46)$$

$$\approx 1333 \text{ x } 0.28$$

$$\approx 373.24$$

Select size as 375 kVAR

One Line Diagram / Single Line Diagram

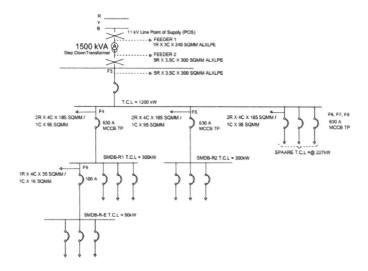

APPENDIX

Abbreviations

- MDB – Main Distribution Board
- SMDB – Sub-Main Distribution Board
- LDB – Lighting Distribution Board
- PDB – Power Distribution Board
- FDB – Fuse Distribution Board
- ACDB – Alternating Current Distribution Board
- DCDB – Direct Current Distribution Board
- MPDB – Main Power Distribution Board
- ELDB – Emergency Lighting Distribution Board
- MCC – Motor Control Centre
- PCC – Power Control Centre
- PMCC – Power and Motor Control Centre
- ATS – Automatic Transfer Switch
- APFCP – Automatic Power Factor Correction Panel
- MCB – Miniature Circuit Breaker
- MCCB – Molded Case Circuit Breaker
- ACB – Air Blast Circuit Breaker
- SF_6 - Sulphur Hexa Fluoride Circuit Breaker
- VCB – Vacuum Circuit Breaker

- ELCB – Earth Leakage Circuit Breaker
- RCCD – Residual Current Circuit Breaker
- ROCB – Residual Over Current Circuit Breaker
- LBS – Load Breaker Switch
- HTSFU – High Tension Switch Fuse Unit
- GCP – Generator Control Panel
- VSS – Voltage Selector Switch
- ASS – Ampere Selector Switch
- AVR – Automatic Voltage Regulator
- AMF – Auto Main Failure
- BDV – Break Down Voltage
- COS – Change Over Switch
- GI – Galvanized Iron
- GLS – General Lighting Services
- HBC – High Breaking Capacity
- HRCF – High Rupturing Capacity Fuse
- IP – Ingress of Protection
- OLR – Over Load Relay
- ELB – Earth Leakage Relay
- LED – Light Emitting Diode
- MSB – Main Switch Board
- NGR – Neutral Ground Resistor
- OCR – Over Current Relay
- OLTC – On Load Tap Changer
- OTI – Oil Temperature Indicator
- RMU – Ring Main Unit

- SWG – Standard Wire Gauge
- SWA – Steel Wire Armoured
- AWA – Aluminium Wire Armoured
- WTI – Winding Temperature Indicator
- VVVF – Variable Voltage &Variable Frequency
- DOF – Drop Out Fuse
- SMFB – Shield Maintenance Free Battery
- JB – Junction Box
- PB – Push Button
- TB – Terminal Box
- LCB – Local Control Board
- LCS – Local Control Station
- NO – Normally Open
- NC – Normally Closed
- ACCL – Automatic Cum Change Over Current Limiter
- AC – Alternating Current
- DC – Direct Current
- A/C – Air Condition
- AFL – Above Floor Level / Above Finished Level
- CCTV – Closed Circuit Television
- EWB – Electric Water Boiler
- FL – Floor Level
- FOC – Fiber Optic Cable
- KHz – Kilo Hertz
- MHz – Mega Hertz

- MS – Mild Steel
- NTS – Not To Scale
- REV – Revision
- WD – Working Drawing
- Φ – Phase
- C – Conduit
- CD – Candela
- DB – Decibel
- EF – Exhaust Fan
- ELEV – Elevator
- ECC – Earth Continuity Conductor
- EM – Emergency
- F – Fuse
- FA – Fire Alarm
- kVA – Kilo Volt Ampere
- kVAR – Kilo Volt Ampere Reactive
- W – Watts
- kW – Kilo Watts
- MW – Mega Watts
- HP – Horse Power

Codes and Standards

- NEC – National Electrical Code
- NC – National Building Code
- NEIS – National Electrical Installation Standards
- NECA – National Electrical Contractors Association
- BS – British Standard
- BEE – Bureau of Energy Efficiency
- IEEE – Institute of Electrical & Electronics Engineering
- IEC – International Electro Technical Commission
- ISO – International Standard Organization
- ISOH – Institute of Occupational Safety & Health
- OSHA – Occupational Safety & Health Administration

ABOUT THE AUTHOR

Japhet K. Lagat's books have received starred reviews in Publishers Weekly, Library Journal, and Booklist.

Before he started writing Electrical Engineering Books, Japhet got a Bachelor's degree in Electrical and Electronics Engineering from Technical University of Mombasa (TUM). After that, just to shake things up, He went to Engineering school at the University of Nairobi and graduated. Then he did a handful of projects with some really important people who are way too dignified to be named here. He is an Electrical Consultant and Projects Manager and writes engineering books and articles.

When he is not writing in his favorite coffee shop, Japhet spends most of his time reading, designing, traveling the world, catching his favorite Broadway shows. An admitted sports fanatic, He feeds his addiction to athlete by watching world championships on Weekend afternoons.

Printed in Great Britain
by Amazon